Zero G Mysteries an(
Weightlessness Attraction

© Copyright 2010
All rights reserved.

ISBN 978-145-36768-4-4

This zero Gravity book and its companion website are undergoing constant improvements. We welcome your comments, suggestions and criticism. We commented on other publications for the sole purpose of learning the subject. We can be contacted at info@0gsuite.com

About the website www.0gsuite.com (0 is the number zero, not to be confused with the letter O)
Some contents are considered not suitable for the book. Few things were deleted from it. Since those topics are also a significant part of the zero G business and human activities in future space exploration, the contents will be moved to the companion website in the near future.

About the author
Jets Hunt is a displaced engineer. His research papers were published in the Global Positioning System and transportation communities. In addition to this book, he has also written GPS Puzzles and the Sherlock Holmes Mystery. He was dumped from an Air Force academy to a dirty, rotten infantry platoon because of a deviated septum in early years. A flying officer candidate must be able to breathe well in high altitudes. He is also doing researches on the impact of technologies on modern military and comparative military history in recent years.

Table of Contents

CHAPTER 1: WHAT IS ZERO GRAVITY? 4

SIMULATING ZERO G ON LAND 8
REDUCED AND VARIABLE GRAVITIES 10
ARTIFICIAL GRAVITIES 10
ZERO GRAVITY EXPLAINED BY DIAGRAMS 12
A SURVEY QUESTION 16

CHAPTER 2: MISCONCEPTIONS ABOUT ZERO GRAVITY 17

NEWTON'S CANNON BALL 24
TEST YOUR PRECONCEIVED NOTIONS ABOUT WEIGHTLESSNESS 30
SOME MYTHS 31
JET FIGHTERS OR JET AIRLINERS 33

CHAPTER 3: AMUSEMENT ATTRACTIONS 35

PARABOLIC ZERO G RIDES 35
VERTICAL ZERO G RIDES 39
IMPLEMENTATIONS 44

CHAPTER 4: THE PHYSICS AND MATH OF 0 G 50

CONTACT VERSUS NON-CONTACT FORCES 50
SCALE READINGS AND THE WEIGHT 53
KINEMATIC EQUATIONS NEEDED IN THE CHAPTER 54
WHAT REALLY HAPPENED IN A ZERO G FLIGHT 54
NEWTON'S CANNON BALL MAGIC 56
DEFINITIONS NEEDED IN THE CHAPTER 58
SIMULATING ZERO G AND REDUCED GRAVITIES WITH ELEVATORS 59
HOW TO SIMULATE VARIABLE GRAVITIES ON A PARABOLIC TRACK 61
INTERESTING FACTS USED BY ZERO GRAVITY 67

APPENDIX 1   GLOSSARY 71

APPENDIX 2   CIRCULAR MOTIONS AND ARTIFICIAL GRAVITIES 75

APPENDIX 3   REFERENCES: 78

Chapter 1: What is Zero Gravity?

Zero gravity, or weightlessness, is a phenomenon experienced by an object with no contact forces acting upon it. Both people on Earth and the weightless astronauts in a spacecraft in the Earth's orbits are constantly being pulled towards the Earth by the gravitational force. The nature of the gravitational force, a non-contact force, is the reason that there are so many misconceptions about zero gravity.

Believe it or not, Wikipedia's and Encyclopædia Britannica's definitions of the word "weightlessness" as "a phenomenon experienced by people during free-fall." and "a condition experienced while in free fall" are only half true incomplete statements. Yes, weightlessness can be created during a free fall. But weightlessness or zero gravity can also be experienced while objects are climbing up. This book attempts to eliminate such misconceptions by discussing some real examples of zero gravity.

For decades, NASA used modified KC-135s (more commonly known as Boeing 707s) maneuvered along a parabola flight path by specially skilled pilots to train astronauts to prepare themselves for the feeling of weightlessness inside an orbiting spacecraft. Russian and European space agencies use modified Ilyushin-76s and Airbuses to train their personnel in a similar fashion. Today, a few companies in the US and other countries, such as Zero G Corporation or bestRussianTour.com, are providing weightless flights as entertainment for paying passengers, with a ticket price of ~USD $5,000. However, the quality of the zero G flights is restricted by the maneuverability of the jetliners and the skill of the pilots, and the weightless time duration is short, twenty to twenty seven seconds for each cycle. Virgin Galactic is offering sub-orbital parabolic flights with a few minutes of weightlessness at an understandably much higher cost. An issue with Virgin Galactic's Space Ship 2 is that it is the size of a small executive jet; there is hardly any room to float while weightless, compared to large commercial jets like Boeing or Airbus. Virgin Galactic's emphasis is space tourism and the view. There are two windows for each passenger. Those thinking ahead are already exploring the clinics and mechanics of and intimacy in zero G.

All of this is just a fantasy for 99.999 percent of us on Earth. The zero G experience is available for most people only in the form of a drop tower or roller coaster rides in an amusement park. But, tying oneself to a seat and experiencing zero G for less than 3 seconds is far too short compared to the sensation of zero G float in space.

The purpose of this book is to investigate the future of a longer, weightless floating alternative at the price of admission into Disney's theme park or a night's stay at a hotel. It will be the next amusement attraction.

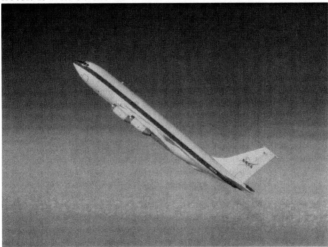

NASA KC-135 performing a zero G flight

An Airbus A-300 performing zero G flight

Inside a zero G flight

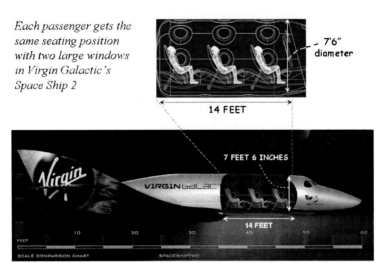

Each passenger gets the same seating position with two large windows in Virgin Galactic's Space Ship 2

Inside a Virgin Galactic's Space Ship 2

Please check the companion website http://www.0gsuite.com for the video of astronauts' weightless life in space

Most of us have enjoyed watching astronauts 'floating' around in the space shuttle on MIR, or now-a-days in the International Space Station (ISS), Why is this possible, and do we have an alternative and cheaper ways to obtain weightlessness other than going into space or zero G flights? The answers are yes, yes and yes. We shall explain more in details regarding why, how and what later in the book.

Simulating zero G on land

Very few people know that everyday processes are performed on Earth in zero or micro gravity, but the concept is fairly old. It is a common practice to cast steel ball bearings or glass marbles by releasing drops of molten steel or glass from the top of a tower. The drops become solid during the fall and harden when submerged in a water bath. The final product is almost perfectly round. Optical fibers are also made with a similar manufacturing technique.

Many people know that there are drop towers and roller coasters in most amusement parks around the world. Thrill seekers have only a transient experience (less than 3 seconds) with zero gravity in a drop tower or rollercoaster rides because the tower structures of this kind would be unstable or too expensive if they are taller than 350 ft. (120 meters). The riders must be tied to the seats for safety measures. It is impossible to float like in a zero G flight or orbiting spacecrafts. These are the limitations of simulating zero G on land today.

 An amusement park drop tower can only use the downward motion to produce zero G.

Fortunately, the advances in the latest technologies from various disciplines in industry are capable of breaking through these limitations and achieving better, longer zero gravity or weightlessness on land than the expensive zero G flights and space flights today. Thanks to the development of high-speed magnetic levitation transportation (maglev for short) and other technologies, it is now possible to achieve zero G using land facilities. These new systems of zero G amusement rides will bring down the cost of weightlessness dramatically, making it possible to offer these rides at the ticket price to an amusement park today. These technologies will be described in Chapter 3.

Reduced and Variable Gravities

Most of us may have already known that there is only 1/6 of the gravitational force on the surface of the Moon as here on the surface of the Earth. There is about 3/8 of the gravitational force on the surface of the Mars. That is to say, a person who weighs 180 lbs (82 kilograms) would weigh only 30 lbs (13.6 kg) on the Moon and 67.5 lbs (30.7 kg) on Mars. In case the readers are wondering why and how these numbers are obtained, the results are based on Newton's laws of motion and universal gravitation. They are not an esoteric subject to learn and will be used in Chapter 4, "The Physics and Math of 0 G". Most of the basic principles behind the topics of this book are based on middle to high school science curriculums.

Reduced gravity means creating gravitational field of strength between zero and the Earth's gravitational field. Simulations of the Martian and lunar gravities are routinely done by commercial zero G flights. The company Zero G Corporation flies parabolic paths designed to offer lunar gravity (one sixth your weight) and Martian gravity (one third your weight) before flying at zero G weightlessness. This is achieved by flying at a larger arc over the top of the parabolic path. This arrangement serves two purposes: to allow passengers to have more diverse experiences and to reduce the possibility of motion sickness during weightlessness. NASA used the nickname "vomit comet" for their zero G flights. Reduced gravities can also be created by means of drop towers. The theory and practices will be discussed in detail in the later chapters.

Artificial Gravities

A spinning spacecraft will be able to produce artificial gravity as in the following diagram. By adjusting the rotational speed, the outer edge of the spacecraft can simulate any gravity desired. This kind of spacecraft design was conceived long time ago, but never used in previous space flights most likely because of engineering tradeoffs. Exercise machines are provided in spacecrafts to reduce the adverse effects of weightlessness for the astronauts. In the future long term space exploration, this kind of man-made gravity spacecraft may be necessary. Please check the Appendix 2 Circular Motions and Artificial Gravities for more information.

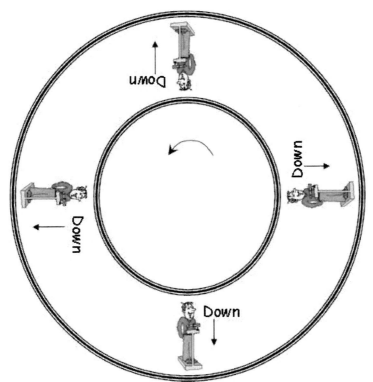

Artificial gravity in a spinning spacecraft

Zero gravity explained by diagrams

Being on the surface of the Earth is equivalent to being in a spacecraft in outer space with a rocket engine thrust. Gravitational acceleration g on the Earth's surface is equivalent to an acceleration a = g in outer space.

When the spacecraft engine is ignited and accelerating, the observer or an instrument inside cannot detect any difference from being on the Earth. Everything is falling down with an acceleration of $g=9.81$ m/s^2. This is called the principle of equivalence which is also the basic principle of Einstein's theory.

If the acceleration is more than g, it is called a high G. For example, 2G means that your weight is doubled. If the acceleration is less than g, it is called reduced G, like on the surface of the Moon or Mars. Zero Gravity or weightlessness is experienced inside a free falling elevator on the surface of the Earth. The observer inside the elevator can not detect any weight.

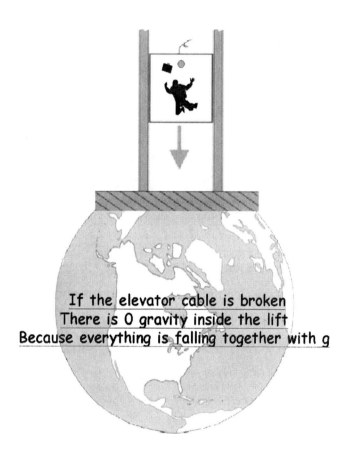

If the elevator cable is broken
There is 0 gravity inside the lift
Because everything is falling together with g

Zero G is also experienced inside an orbiting spacecraft because it is falling forever along the orbit in the same way the Moon is falling but not crashing into the Earth, as Newton's apple did.

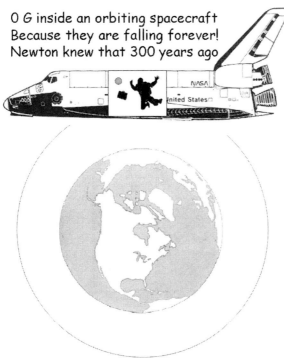

0 G inside an orbiting spacecraft
Because they are falling forever!
Newton knew that 300 years ago

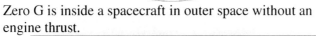

Zero G is inside a spacecraft in outer space without an engine thrust.

A survey question

It is time to make a survey questionnaire. We encourage the reader to ask the same question to others.
Do you think zero gravity/weightlessness can be in an elevator while it is climbing upwards? Why or why not?

Creating 0 G while the elevator is moving up ?

Clue: A Google image search for "zero G" will result in pictures of aircrafts climbing up like the first two diagrams in this chapter. Why?

Chapter 2: Misconceptions about Zero Gravity

Let us investigate the following five diagrams. They are designed by the experts in the fields of Zero G business. Can you find any inconsistencies?

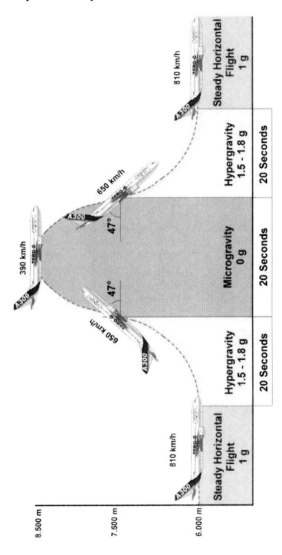

Parabelflug

Injection point   Recovery point

8500 m

7600 m   47°   42°   370 km/h

6100 m   570 km/h

825 km/h

Hyper-Schwerkraft	Schwerelosigkeit	Hyper-Schwerkraft
20 Sekunden	22 Sekunden	20 Sekunden
	1 Minute 10 Sekunden	

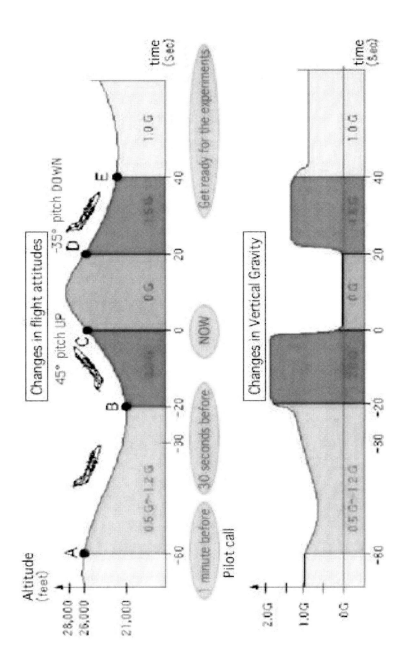

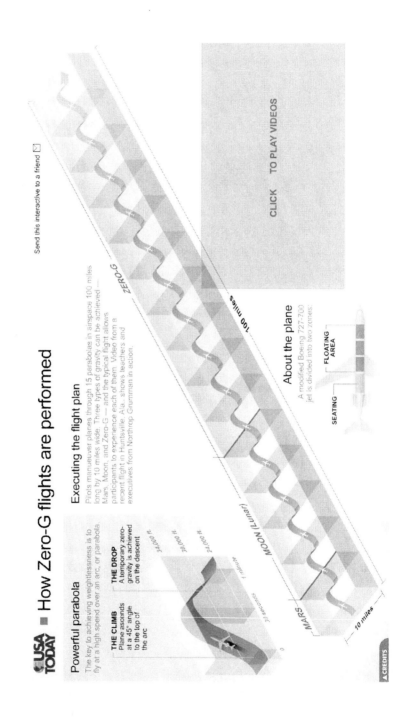

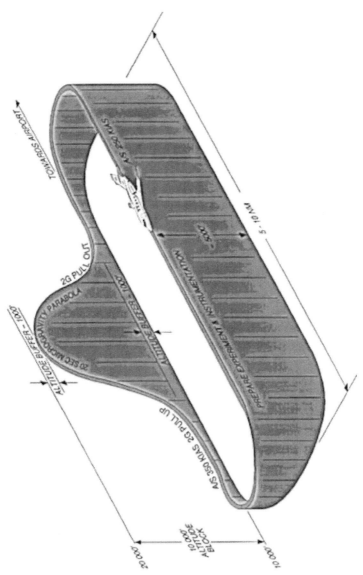

The baseline distance is 5 to 10 Nautical Miles NM (1 Nautical Mile NM =1.15 mile, 1 NM = 1.852 Kilometers) in this diagram

We shall answer and explain in a little while for readers to have more time to examine the diagrams and think about what is weightlessness or zero gravity.

It is not surprising that many people have wrong ideas about weightlessness and zero gravity such as the following:
Astronauts inside an orbiting spacecraft are *weightless*

because space is a vacuum and there is no gravity in a vacuum.

because space is a vacuum and there is no air resistance in a vacuum.

because the astronauts are at a location so far from Earth's surface that the the Earth's gravitation force is negligible.

because there is no gravity in space, so they do not weigh anything.

Some people believe that weightlessness is due to the absence of air in space. This misconception lies in the idea that there is no force of gravity when there is no air. According to this notion, gravity does not exist in a vacuum. But this is not the case. Gravity is a force which acts between the Earth's mass and the mass of other objects which surround it. The force of gravity can act across large distances to the Moon and its affect can even penetrate across and into the vacuum of outer space. People with this misconception are confusing the force of gravity with air pressure. Air pressure is the result of surrounding air particles pressing upon the surface of an object in equal amounts from all directions. The force of gravity is not affected by air pressure. While air pressure reduces to zero in a location void of air (such as space), the force of gravity does not. Indeed the vacuum results in the absence of air resistance; but this would not account for the weightless sensation.

We can never create weightlessness in a tank on the Earth, even in a vacuum chamber. In fact, the astronauts working at the International Space Station (ISS) are not working in a vacuum (people would be killed violently in a vacuum), but under normal atmospheric pressure conditions. It is not possible to shield against gravity with a vacuum. Having no air resistance and weightlessness are different things; they not related to each other whatsoever.

The astronauts are never so far away from the Earth that the gravitational pull can be negligible. The gravitational force will be reduced by only a few percent in Earth's orbit.

So, why is gravity not present in a spacecraft? Actually it is. Astronauts, the spacecraft, and all the equipment are doing nothing but constantly falling towards the Earth - they are constantly free falling. The reason that the spacecraft is not crashing into the Earth is that it is moving at sufficient horizontal speed, approximately 7.5 km/s in low earth orbit (LEO is 300-600 km above the ground), so that the spacecraft is not falling vertically, but it follows a curved course (known as an orbit) around the Earth. This velocity is obtained from the rocket engines during launch into orbit. We can apply the same idea to the Moon. The Moon is moving at about 1 km/s along its orbit around the Earth.

Since everything is falling at the same speed, the crew members onboard the spacecraft experience weightlessness. If the astronauts stands on a scale it would read 0, but an observer far away would realize that the reason for this is that the astronaut and the scale are both following each other in a free fall along the orbit. So yes, the astronaut is weightless because there is no contact force between him/her and the scale, but they are still affected by gravity!

Newton's Cannon Ball

The above facts were first discovered by Sir Isaac Newton. He predicted the creation of artificial satellites and man-made moons almost three hundred years before the Russian Sputnik, the first man-made satellite in the world in the 1950's.

Newton's reasoning was relatively simple. He imagined an extremely powerful cannon on top of a high mountain firing cannon balls horizontally as in the following diagram. His conceptual experiment was to shoot cannon balls faster and faster by ignoring the air resistance upon the cannon balls. Projectiles A and B fall back to earth because they do not have enough speed to free-fall along with the curvature of the Earth. One of these cannon balls will finally be fast enough to fall forever along with the Earth's surface and achieve circular orbit, as projectile C does in the diagram. That is exactly the same reason the Moon is not falling down to the Earth and the reason an apple fell onto Newton's head while he was dreaming under an apple tree. The Moon has a large horizontal speed but the tree top apple does not. We'll be able to figure out how fast the Newton's cannon ball or apple should be moving horizontally in order to stay in orbit in Chapter 4. The Moon's horizontal speed can be derived from the perimeter length of the Moon's orbit divided by the period of the Moon.

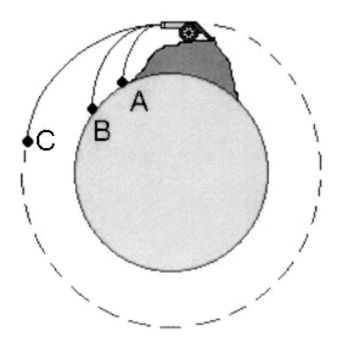

Newton's cannon ball is as good as his apple in science

Unfortunately, the wonderful story above leads to yet another common misconception among the scientific intellectuals like those who organized the weightlessness articles for Encyclopædia Britannica and Wikipedia, and the artist of the fourth diagram of zero G flight in the beginning of this chapter. Many people who learn this definition of free fall believe that "weightlessness is a phenomenon experienced by people during free-fall." This is only half true because weightlessness will also be experienced during "free-climb", "free ascending" or free ascension".

The author was an infantry ordnance officer during the Cold War. We had plenty of firing maneuvers during those days. Mortars and howitzers are fired with a high angle of elevation similar to the zero G flight climb angles. With good eyesight, we could actually see the relatively slow mortar shells shooting out of the muzzle, climbing to the top and then falling downward along a parabolic projectile path. Now, let us learn something from Sir Isaac Newton's science method by imagining if there was an ant inside a flying empty shell. What would happen to the ant during the flight of more than 20 seconds for the US-made 4.2 inch and 81mm mortars shells?

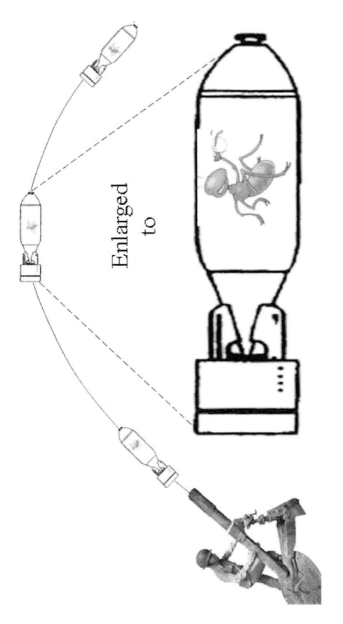

Imagine that, there was an ant inside my empty mortar shell! How would he be doing during the 20 second flight?

An 81mm Mortar in Action

After the shell leaves the barrel, there will be no external contact force acting on the shell and everything inside. Of course, we are ignoring air drag during our mental exploration just like Newton. The shell and everything inside is in a free ascending, free climbing or free flying state. They are following each other along a parabolic flight path up because the shell has a very large initial speed upward. The shell and everything inside are slowing down at the same rate of $g=9.81$ m/s^2 while climbing to the top. During this period, the ant inside will be in a weightlessness state because the inner shell wall will not push nor pull on him. He has no sense of what is up or down during this time. An observer from far away would realize that the reason for this is that the shell, the ant and everything inside are all following each other in a free climbing, and then free falling parabolic path. There is no contact force acting between them theoretically. Please note that we are assuming ideal conditions, as necessary in many experiments. However, factors like side wind and air drag must be taken into account by the operators of the artillery to fire at pinpoint accuracy. Training and coordination among the gunners are important in a real situation as seen in the above photos.

The results of this conceptual experiment are exactly like the first three and the fifth diagrams of the zero G flight at beginning of the chapter. The climbing of the aircraft accounts for half of the zero G duration. The other half comes from the aircraft diving or free falling. Again, if a passenger in the aircraft stands on a scale during this period, it would read 0, but an observer far away would realize that the reason for this is that the passenger and the scale are both following each other in a free climbing and free falling parabolic path. There is no force between the two. Free climbing and free falling is actually the same thing, in both cases we are accelerating down with gravity, only the direction of travel velocity is different, and has the same zero G effects inside the free flying object. The same zero G effect exists inside the freely flying object.

The mistake of the fourth diagram is that it does not indicate that a temporary state of zero G is also achieved during the climb to the top of the arc. The misfortune Boeing 727 called G-Force One doing a free falling nose dive for 30 seconds as in the diagram would be in an unrecoverable condition. We shall also find out that a 30 second free-fall dive would need a height of at least 14,456 ft or 4,400 meters in Chapter 4.

Similar to the art of artillery, the quality of weightlessness on today's zero G flight is largely dependant on the skill of the pilots. During the zero G duration time, engine power is only used to compensate for the air drag in order to keep the aircraft on an ideal parabolic path and at the speeds needed along the path. Many factors will affect the zero G flight, such as the performance and maneuverability of a particular aircraft, the weather, speed, acceleration and pitch control etc. This kind of delicate maneuvering cannot be done with a smart autopilot system on board yet.

Another common misconception about zero G is that it is possible to create weightlessness experiments in a deep diving pool. This is not true, since everything in the pool is affected by gravity, but the buoyancy of the water can be used to reduce most of the weight of the astronauts, which is useful for training astronauts in extra vehicular activities (EVAs, outside of the spacecraft). It is, however, important to note that the organs of the astronaut are still subject to gravity while working in a diving pool. Also the astronaut can move around in the water by swimming, something which is not possible in space.

After discussing these causes of zero G and weightlessness, we encourage the readers to conduct the following questionnaire to other people to determine that this concept is actually not intuitive.

Test your preconceived notions about weightlessness

Is it possible to create weightlessness in an elevator while it is moving up? Answer: Yes, if the elevator started with a large upward speed. It is moving up while slowing down at the rate of g, the gravitational acceleration. The passengers inside will be weightless.

Please note that weightless is not the same as having no mass in science.

Astronauts on the orbiting space station are *weightless* because...

a. there is no gravity in space and they do not weigh anything.
b. space is a vacuum and there is no gravity in a vacuum.

c. space is a vacuum and there is no air resistance in a vacuum.

d. the astronauts are far from Earth's surface at a location where gravitation has a minimal affect.

Most people will be confused about the notion of having no weight. The cause of weightlessness is quite simple to understand. However, the stubbornness of one's preconceptions on the topic often stands in the way of one's ability to understand. We conducted similar survey questionnaires over the years and will distribute a similar survey on the website. The results we collected were interesting.

Some Myths

Newton's monumental achievements like the equation $F = ma$ etc. are easy to remember but also can be very intriguing to interpret. There were indications suggesting that Newton himself was embarrassed by his invention of the non-contact or acting-at-a-distance forces such as the gravitational pull reaching out through space to infinity. Newton's apple, the most famous legend in science was told by an older Newton to his biographer William Stukeley. At the age of 21 in August 1665, Newton went back home because his university was temporarily closed as a precaution against the great plague spreading across Europe. Newton's mother had sent him to the university because he was too lazy to be a decent farmer. Newton's private studies at home over the next two years resulted in the development of calculus and the law of gravitational forces. He had kept the story and theory of the apple to himself for a long time until he told his first biographer years later. It is possible that he was uneasy about the fact that he might be laughed at and ridiculed by his classmates and professors if they had learned of the invisible, untouchable forces of gravity radiating through the space of his wild imagination.

For thousands of years before Galileo and Newton, people did not know the difference between weight and mass. Weight and mass was used interchangeably then. After Newton, the scientists knew the difference between them. Weight is caused by a force. The mass of an object is a fundamental property of matter; it is a numerical measure of an object's inertia and a measurement of the amount of matter in the object. Definitions of mass often seem circular because it is such a fundamental quantity that it is hard to define in terms of something else.

In our everyday usage, *mass* is still taken to mean weight after all these years, but in scientific use, they refer to different things. This is why many people are still confused. For example, a bathroom scale having both pound and kilogram readings will show that 1 kg equals 2.2 pounds. But, wait, kilogram is a unit of mass and pound is a British unit of force. So are we comparing apples with oranges? The answer is "Yes, we are indeed equating different things!" We have also done so earlier in Chapter 1 of this book. The scientifically precise way of using the above equation is by multiplying the gravitational acceleration at the surface of Earth, g=9.8m per second squared, by 1kg. That is, 1kg*9.8m/sec/sec=2.2 lbs. Since g is constant everywhere we can go, we ignore it in many everyday applications. However, this means that 1kg is no longer equal to 2.2 pounds on the surface of the Moon. It is only true near the Earth's surface, which is really good enough for most everyday applications.

Jet Fighters or Jet Airliners

Let us go back to the first few diagrams and examine the details again. They are somewhat different. For example, the first one has 20 seconds of zero G and the second one has 22 seconds. The third one has 20 seconds, and the fourth one has almost 25 seconds using a Boeing 727-700 jet. The elevation and diving angles at the injection and recovery point are also slightly different. They are 47, 47, 45 and 45 degrees upwards and 47, 42, 35, 45 degrees downwards, respectively. These minor discrepancies are caused by the different types of aircrafts used. Since they are all comparable jetliners, their performance characteristics are similar. Now, what if we used a jet fighter, a much more maneuverable jet? Will that make the 0 G duration longer or shorter? Most of our questionnaire answers are correct. But only some of the answers provided an explanation.

An F-16 Falcon fighter jet

An F-16 can achieve almost one minute of zero-G because it has a much higher (supersonic) initial, upward speed upward. The reason fighter jets are not good for a fun 0 G flight is because of the payload. There is no cabin space to float either, so there is no room to move around for fun. The passenger has to be tied to the seat as in a drop tower at amusement parks.

Chapter 3: Amusement Attractions

The purpose of this book is to investigate the future of achieving better and longer zero G weightlessness for a cost of $50 instead of $5,000. This business is not meant to be developed by a space agency, nor the aerospace companies. Rather, companies like Disney, Six Flags or Hilton hotels would run this new kind of amusement attraction. The technology already exists. Just as owning a cell phone or PC and using the Internet were only dream for the general population not very long ago, the goal is to bring the cost down so that everybody can afford and enjoy it.

Parabolic Zero G rides

People have already created transient weightlessness on land by drop towers and roller coasters much earlier than space and zero G flights. There are two problems that prevent engineers from designing a longer zero G ride on land. One is the structure of the tower and roller coaster. They cannot exceed a height of more than 400 feet (130 meters) or else it would be prohibitively expensive or unstable in bad weather. Vertical height is the physical limitation of the zero G time. The other problem is that roller coaster carts cannot reach the speed necessary to perform a long zero G ride like an aircraft can.

Actually, the second problem facing us is already solved. Modern High Speed Rail (HSR) trains are fast enough to compete with a slow jet. What is preventing high-speed trains from getting even faster today is not limited technology. The reason is economic; air drag will waste most of the engine power in high speeds. Aircraft flying at high altitudes enjoy much less air drag because the air is much thinner up there.

A **vactrain** is a proposed design for future high-speed rail transportation by building maglev lines through evacuated (vacuum, air-less) or partly evacuated tubes or tunnels. The technology is currently being investigated for development of regional networks in Japan and under research in several institutions around the world. Advocates have suggested establishing vactrains for transcontinental routes to form a global network. Eliminating air drag could permit vactrains to use little power and to move at extremely high speeds, up to 4000-5000 mph (6400–8000 km/h), 5-6 times the speed of sound at sea level. For the purpose of achieving weightlessness in a zero G parabolic path, the speed of sound (768 miles per hour) would be more than enough.

Let us do an analysis of the parabolic paths needed for zero G flights. From the first few diagrams in Chapter 2, we found that the height of the curved path is 8,200 feet (about 2,500 meters), and the baseline distance is about 6.7 miles (10 kilometers). Ten miles is within an intercity regional transportation range. The Japanese maglev test track in Yamanashi County is more than 10 miles. The length of the maglev test facility in Emsland, Germany is 20 miles long. The length of English Channel tunnel is more than 30 miles.

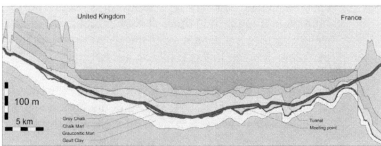

English Channel Tunnel

The 8,200 feet vertical height needed for weightlessness zero G drop would be impossible for any human-made structure like bridges or towers. However, a tunnel is a structure supported by the Earth itself. The English Channel tunnel goes down and up across the Channel. A zero g track path would need to go up and down as the following diagram.

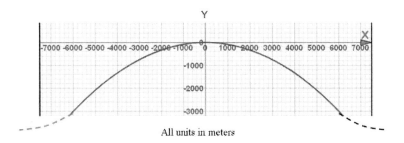

All units in meters

Many mountain terrains around the world can cover the whole length of the parabolic path in a tunnel needed for the weightless rides. The tunnel is needed if the air is to be partially evacuated for the speed and economy of the vehicle system. A pair of locking gates is placed at the entrance and the exit of the tunnel in order to keep the low air pressure inside the tunnel. Each pair of the locking gates forms a chamber for the vehicle to park while waiting for the air pressure inside the chamber to adjust. The function of the locking gates for different air pressure levels is similar to the water locking gates used in Panama Canal for different water levels. The passenger cabin of the vehicle is pressurized like the cabin of an aircraft. The ideal location of this kind of tourist attraction will be near populated metropolitan areas near mountain ranges with a population in need of fast regional transit.

The system is based on mountain terrains and existing technologies such as Maglev high-speed trains, pressurized passenger cabins, tunnel boring technology, rail-based transportation, vacuum technology, etc. This system is economically feasible and can be integrated into an intercity passenger transit system.

The advantages of this kind of zero G rides over today's zero G flights are obvious. Zero G track vehicles can be automated, without a pilot or concern for the weather. Land vehicles can be dispatched much more frequently with more passengers and heavy payloads. The cost of operation and maintenance will be low. All these factors will make a $50 weightless ride a reality.

The critics of this system are skeptical because a vactrain project of this kind is still quite an undertaking and too futuristic. Knowing the business model can be operated 24/7 as described in the complementary website www.0gSuite.com (0 is the number zero, not to be confused with the letter O), building a test track of fifteen miles at a right location will be itself a lucrative business according to the business analyses from independent studies. It will not like the showcase demo test tracks in Japan, Germany and Korea idling most of the time. The operations will never need to be subsidized by the government. It will need to be initiated by a public and private partnership though. The project proposal is also in accordance with Obama administration's vision of high speed rail in USA. We don't have to follow. We can frog leap ahead of European and Asian nations in High Speed Rail HSR transportation.

Today's commercial zero G flights can simulate reduced gravities like on the Moon or Mars by taking different parabolic flight paths as we mentioned before. The proposed zero G high speed train will be on a fixed parabolic track. Is it impossible for a land vehicle to do the same? Let us conduct a survey questionnaire here again. What do you think and why? The answer will be in the last chapter of the book, the physics and math of zero gravity, because there will be some mathematical concept involved and we encourage the serious readers to think about it. It can be quite a conceptually challenging puzzle.

Vertical Zero G rides

Roller coasters and vertical drop towers are two kinds of facilities to create weightlessness on land.
Mathematically, the vertical drop tower is a special case of the roller coaster where the horizontal speed is zero. The parabolic zero G high speed train ride mentioned above is like a huge roller coaster. Let us investigate the alternative here. The drop tower is a kind of elevator. We have used an imaginary vertical elevator in chapter one to explain weightlessness. Is it possible to build a large and very high elevator with enough room to float around and a long enough zero G duration? If such an elevator is to be built, it would also be much better than today's drop towers and the zero G flights. We need to investigate the factors such as the available and necessary height, speed, and acceleration as we did before.

How difficult is to acquire a vertical height of more than 1,000 meters? Of course, a man-made tower structure above ground exposed to the weather is not an option for the reasons we have discussed before. Fortunately, there are many places on the Earth itself that can provide such a structure. As a matter of fact, some of the mine pits are more than that deep. Mountain terrains are also possible. Modern skyscrapers are few hundred meters high, and future skyscraper will be even higher. The combination of all the possibilities above will be discussed.

If a 1,000+ meter high elevator is to be built, the cable, pulley and motor technology for the conventional elevators would not be good enough. A new type of elevator power system is needed.

A linear electromagnetic motor or linear motor is basically an electric motor that has had its stator "unrolled" so that instead of producing a torque (rotational force) it produces a linear driving force along its length.

There are many designs have been put forward for linear motors. They fall into two major categories, low-acceleration and high-acceleration linear motors. Low-acceleration linear motors are suitable for maglev trains and other ground-based transportation applications. High-acceleration linear motors are normally quite short, and are designed to accelerate an object up to a very high speed and then release the object, such as in roller coasters, catapults or mass drivers and rail guns. They are usually used for studies of hypervelocity collisions, as weapons, or as mass drivers for spacecraft propulsion.

An application of the linear motor is the navy's latest catapult system for aircraft carriers. Conventional aircraft carrier catapults are powered by steam and pistons. There are several drawbacks to the steam catapults, and the navy prefers electromagnetic catapults for the next generation of Gerald Ford Class carriers. Electromagnetic catapults use a feedback control system to control the launch with great precision, which inflicts less stress on the aircraft and allows the catapult to launch a variety of aircraft, from heavy radar surveillance planes to fighter bomber jets to light unmanned aircrafts. All kinds of linear motors can use a feedback control to regulate outputs precisely. A similar electromagnetic motor will drive the power system for the zero G elevators. High quality weightlessness needs the power and the precise speed control of linear motors.

How this kind of zero G elevators will work and how they will be implemented are illustrated in the next few diagrams. Starting from the bottom (Step 1), the elevator is accelerating up and gaining speed. It is to be let go at the injection point. An analogy is the mortar shell accelerating in the barrel tube which is released at the muzzle. During the time from the injection point to the top (Step 2), weightlessness is inside the elevator because everything is free climbing and slowing down with the same deceleration g. After the injection point, only minimum power is needed to keep the elevator at the ideal free climb speeds all the way to the top.

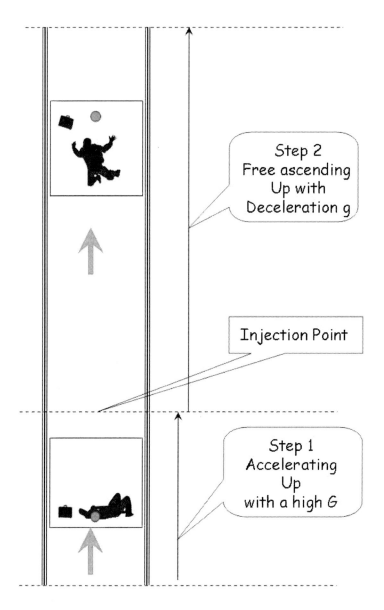

Note that only part of the available height can be used for weightlessness, up to 30% of the total height will be needed for accelerating up and gradual slowing down of the elevator. The injection point is where the elevator will be released for free climbing. The passengers will experience an exciting sensation on the transition from a high G to weightlessness at the injection point and start to float. The recovery point in the next diagram is where the elevator starts to slow down.

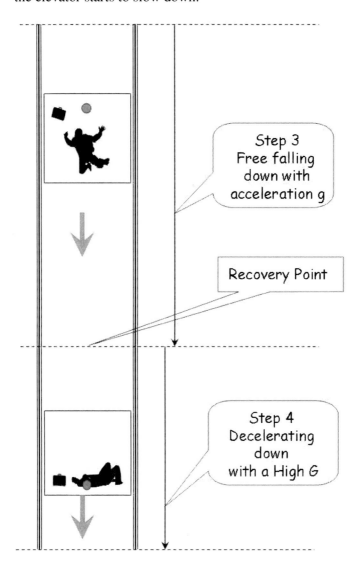

After reaching the top, the elevator is let go for free fall and accelerates downward (Step 3) until reaching recovery point. An analogy is the mortar shell or the zero G jet that starts diving after reaching the top of the arc. During this time, weightlessness is inside the elevator and minimum power is needed to overcome the air resistance and keep the ideal speeds for free falling along the way. After the recovery point (step 4), the elevator starts to slow down with a safe high G to the bottom. After step 4, the whole process from steps 1 to 4 can be repeated as in a parabolic zero G flight for the next cycle of weightlessness ride.

Gradual increase of the gravity is necessary after passing the recovery point to avoid a sudden change from zero G to a high G for the safety of the passengers. We shall investigate how this can be done in the last chapter with some concepts in math. Since there is no safety concern, no provision of procedure is needed for the sudden transition from high G to zero G at the injection point because it is a desired, exciting effect and has no safety concern. However, the high G in Step 1 needs to stay below 4 G's.

Implementations

There are several ways to implement such zero G elevators.

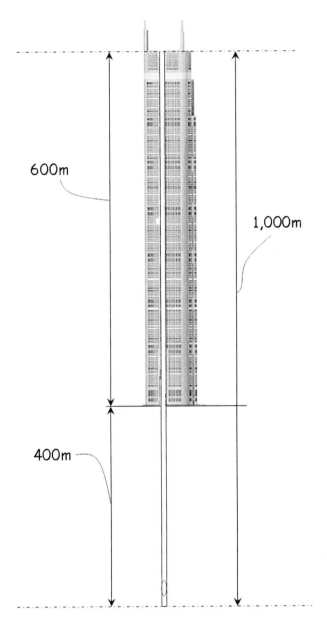

A skyscraper of 600 meters built above a 400 meter deep underground well allows for a total height of 1,000 meters which can produce 25 seconds of weightlessness on each cycle. The Burj Khalifa Building in Dubai is more than 800 meters in height. Future skyscrapers can be even higher.

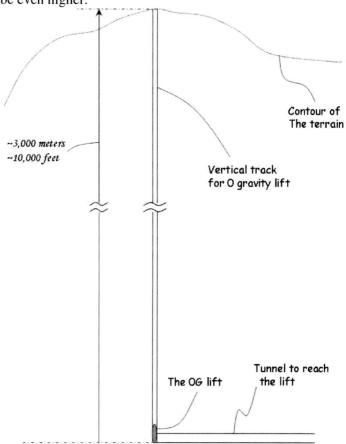

A terrain with a total height of 3,000 meters as the structure for the zero G elevator (lift) can produce ~50 seconds of weightlessness on each cycle.

In theory, no power is needed during free falling and free ascending weightless stages. In practice, however, precisely regulated speed and power output control is always needed to overcome the effects of the air drag and friction forces in order to maintain the ideal free falling and free ascending speeds to simulate zero G weightlessness inside the elevator. The computer is also needed to gradually control the deceleration of the elevator after the recovery point to avoid sudden change from zero G to a high G for the safety of the riders. A sudden transition from high G to zero G is good because it is a desired effect and has no safety concern.

A possible elevator interior design is shown in the following diagram. A multi-deck interior elevator maximizes the passenger payload and the space needed to float during weightlessness. The cabin is pressurized to keep a constant and comfortable environment during the rapid change in altitude for the operation of the system.

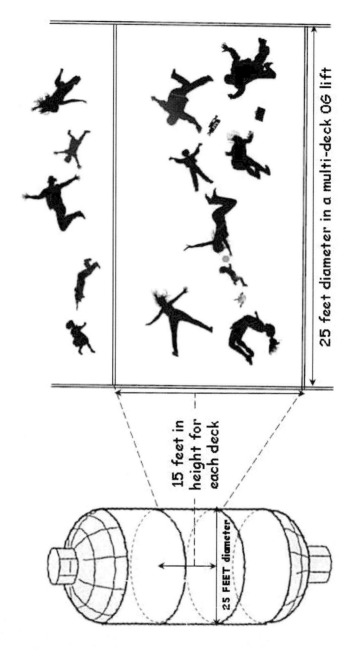

The zero G elevator is based on the new uses of application of existing technologies. We can apply zero

G flight principles to a system of buildings and terrains that support the structure of an elevator and a vertical linear electromagnetic motor mass driver similar to the horizontal electromagnetic catapult to launch aircrafts to be deployed on aircraft carriers. These technologies can be readily modified or adjusted in order to suit the needs to create a zero G elevator.

The advantages of this kind of vertical zero G rides over the previously discussed parabolic zero G path are:
1. It can be located in the mid of a population center and enhances the function of the building as a tourist attraction.
2. The speed needed is much lower than the parabolic rides because the horizontal component of the velocity is 0. By eliminating the need of a partial vacuum, the cost of constructing such facility will be much lower. A few business models utilizing this kind of facility will be discussed in the companion website.

How can the zero G elevators be used to simulate reduced gravities like the gravities on the surface of the Moon or Mars? This is an interesting, easy enough problem for those with a high school physics background. Hardly any mathematics will be involved. We shall explain it in the next chapter. Meanwhile, we encourage the readers to give some thought about it.

Chapter 4: The Physics and Math of 0 G

As quoted from the website Physics Classroom, it is the case for many topics in physics, a conceptual science, that some unlearning must first be done before doing the learning. To put it another way: it is not what one does not know that makes learning difficult; it is what one already knows which makes learning a difficult task. So if the learner has preconceived ideas about what weightlessness is, the learner needs to be aware of this fact.

Contact versus Non-Contact Forces

We mentioned contact and non-contact forces at the very beginning of the book and we should deliberate some more about this concept. Before understanding weightlessness, we will have to tell contact forces from action-at-a-distance forces. Examples of contact forces are push, pull, punch, drag etc. Non-contact forces are magnetic, electricity and gravitational attractions.

As you sit in a chair, you experience two forces - the force of the Earth's gravitational field pulling you downwards toward the Earth and the force of the chair pushing you upwards. The upward chair force is sometimes referred to as a normal force and results from the contact between the chair top and your bottom. This normal force is categorized as a contact force. Contact forces can only result from the actual touching of the two interacting objects - in this case, the chair and you.

The force of gravity acting upon your body is not a contact force; it is often categorized as an action-at-a-distance force. The force of gravity is the result of your center of mass and the Earth's center of mass exerting a mutual pull on each other; this force would even exist if you were not in contact with the Earth. The force of gravity does not require that the two interacting objects (your body and the Earth) make physical contact; it can act over a distance through space. Since the force of gravity is not a contact force, it cannot be felt through contact. You can never feel the force of gravity pulling upon your body in the same way that you would feel a contact force. If you slide across the asphalt tennis court (not recommended), you would feel the force of friction (a contact force). If you are pushed by a bully in the hallway, you would feel the applied force (a contact force). If you swung from a rope in gym class, you would feel the tension force (a contact force). If you sit in your chair, you feel the normal force (a contact force). But if you are jumping on a trampoline, even while moving through the air, you do not feel the Earth pulling upon you with a force of gravity (an action-at-a-distance force). The force of gravity can never be felt. Yet those forces which result from contact can be felt. And in the case of sitting in your chair, you can feel the chair force; and it is this force which provides you with a sensation of weight. Since the upward normal force would equal the downward force of gravity when at rest, the strength of this normal force gives one a measure of the amount of gravitational pull. If there were no upward normal force acting upon your body, you would not have any sensation of your weight. Without the contact force (the normal force), there is no means of feeling the non-contact force (the force of gravity).

Weightlessness is a sensation experienced by a person when there are no external objects touching one's body and exerting a push or pull upon the person. Weightless sensations exist when all contact forces are removed. These sensations are common to any situation in which you are momentarily (or perpetually) in a state of free fall. When in free fall, the only force acting upon your body is the force of gravity - a non-contact force. Since the force of gravity cannot be felt without any other opposing forces, you would have no sensation of it. You would feel weightless when in a state of free fall.

These feelings of weightlessness are common at amusement parks for riders of roller coasters and other rides in which riders are momentarily airborne and lifted out of their seats. However, the duration of the feeling is less than 3 seconds. It is hard to have the sensation of being weightless like the astronauts.

Suppose that you were lifted in your chair to the top of a very high tower and then your chair was suddenly dropped. As you and your chair fall towards the ground, you both accelerate at the same rate of $g = 9.8$ meters per second squared. Since the chair is unstable, falling at the same rate as you, it is unable to push upon you. Normal forces only result from contact with stable, supporting surfaces. The force of gravity is the only force acting upon your body. There are no external objects touching your body and exerting a force. As such, you would experience a weightless sensation. Your weight is as much as you always do (or as little), yet you would not have any sensation of this weight.

Weightlessness is only a sensation; it does not in reality correspond to an individual who has lost weight. As you are free falling on a roller coaster ride, you have not momentarily lost your weight. Weightlessness has very little to do with weight; rather, it is determined by the presence or absence of contact forces. If by "weight" we are referring to the force of gravitational attraction to the Earth, a free-falling person has not "lost their weight;" they are still experiencing the Earth's gravitational attraction. The confusion of a person's actual weight with one's feeling of weight is the source of many misconceptions.

Scale Readings and the Weight

Technically speaking, a scale does not measure one's weight. The scale reading is actually a measure of the upward force applied by the scale to balance the downward force of gravity acting upon an object. When an object is in a state of equilibrium (either at rest or in motion at constant speed), these two forces are balanced. The upward force of the scale acting on the person equals the downward pull of gravity (also known as weight). And in this instance, the scale reading (which is a measure of the upward force) equals the weight of the person. However, if you stand on the scale and bounce up and down, the scale reading undergoes a rapid change. As you undergo this bouncing motion, your body is accelerating. During the acceleration periods, the upward force of the scale is changing. And as such, the scale reading is changing. Is your weight changing? Absolutely not! You weigh as much (or as little) as you always do. The scale reading is changing. But remember: the SCALE DOES NOT MEASURE YOUR WEIGHT. The scale is only measuring the external contact force which is being applied to your body.

Kinematic Equations Needed in the Chapter

We are going to need the following basic equations in the following discussions.

$$v_f = v_o + at$$

$$v_f^2 = v_o^2 + 2as$$

$$s = v_o t + \frac{1}{2}at^2$$

v_f is the final velocity, v_0 the initial velocity, a is acceleration and s the distance traveled, t is the time elapsed. a is the gravitational acceleration $g = 9.81 m/s^2$ in this case. The equations will be simplified if there is no initial velocity to start with or the final velocity is zero. For example, the last equation will be $s = \frac{1}{2} g t^2$, if the object in motion starts from a zero speed.

This is a very useful little equation. Two examples are given below. Let us apply this equation we just learned.

What really happened in a zero G flight

A zoomed-in version of the diagram of Chapter 2 is shown here.

Powerful parabola

The key to achieving weightlessness is to fly at a high speed over an arc, or parabola.

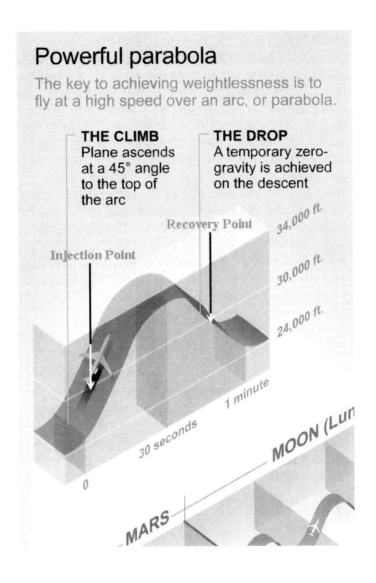

THE CLIMB
Plane ascends at a 45° angle to the top of the arc

THE DROP
A temporary zero-gravity is achieved on the descent

The diagram indicates that zero G or weightlessness time is the time for the vertical drop. The time axis shows that the zero gravity time is from 30 seconds to 1 minute, an elapse time of 30 seconds. By using the equation $s = \frac{1}{2} a\, t^2$ above, where s is the height, a is $9.81\ m/s^2$ and time is 30 seconds, we get a vertical drop of 4,410 meters

(14,465 feet). A 10,000 ft. vertical drop would be far from reality because more than 14,465 ft. is needed for a gradual pull up to a level flight. What really happens is that the pilots engage full throttle, and the aircraft is climbing up until the injection point as indicated in the diagram. There is a high G before the injection point. With a large upward speed accumulated, the pilots then reduce throttle at the injection point to keep the jet on a free ascending parabolic flight path, over the top and dive along the parabolic path to the recovery point. The cabin and its contents experience zero G during this time period. Starting from the recovery point, the pilots pull up the aircraft to start a level flight at 24,000 ft and begin the next climb. The parabolic flight path is needed because we do not want the aircraft to crash into the ground like a Kamikaze.

Newton's Cannon Ball Magic

Another example of using $s = \frac{1}{2} g t^2$ is to figure out how fast Newton's cannon ball needs to be in order to put it in an orbit. The cannon ball falling for 1 second will travel a vertical distance of $0.5 \times 10 \times 1^2 = 5$ meters. We approximate $g = 10$ meters/sec^2. Now, we need to find out how far the cannon ball should travel horizontally in 1 second assuming that the Earth's curvature is about 5 meters.

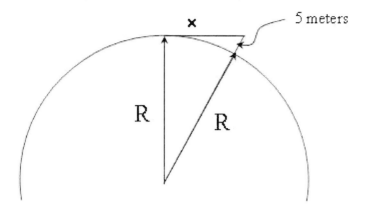

We shall use R=6,371,000 meters as the Earth's radius. By applying Pythagoras theorem on the above diagram, we find that x=8,000 meters. For every 8000 meters traveled along the horizon of the earth, the earth's surface curves downward by approximately 5 meters. So if you were to look out horizontally along the horizon of the Earth for 8000 meters, you would observe that the Earth curves downwards below this straight-line path a distance of 5 meters.

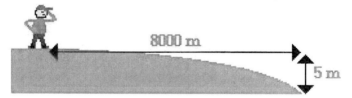

For every 8000 meters along the horizon, the earth curves downward by 5 meters.

That is to say the cannon ball needs to travel 8km per second in order to enter and stay in a circular orbit along the Earth's surface. For any projectile to orbit the earth, it must travel horizontally a distance of 8000 meters for every 5 meters of vertical fall. It so happens that the vertical distance which a horizontally launched projectile would fall in its first second is approximately 5 meters

($0.5 \times g \times t^2$). Newton's magic is not so difficult for us to understand after all.

Definitions Needed in the Chapter

The quantities of distance, velocity and acceleration are directional. The convention is to use a little arrow on top of the variable to denote a vector. Time with unit of seconds is a non-directional quantity. We define velocity is the rate of change of distance as

$$\vec{v} = \frac{d\vec{s}}{dt}$$

The acceleration is defined as the rate of change of velocity

$$\vec{a} = \frac{d\vec{v}}{dt}$$

where

$\vec{a}$ is acceleration in meters per second squared,

$\vec{v}$ is velocity in meters per second,

$\vec{s}$ is position in meters,

t is time in seconds

We need to define one more quantity for our study of zero G elevator operation. Jerk, jolt or surge is defined as the rate of change of acceleration like the following:

$$\vec{j} = \frac{d\vec{a}}{dt} = \frac{d^2\vec{v}}{dt^2} = \frac{d^3\vec{s}}{dt^3}$$

The unit of jerk is meters per second cubed. It will be needed to describe the motion of the 0 G elevator after passing the recovery point. Jerk is used in engineering design for building roller coasters and drop towers. Some precision is necessary for fragile objects such as passengers, who need time to sense stress changes and adjust their muscle tension.

Simulating zero G and reduced gravities with elevators

By now, we should have no doubt that weightlessness can be simulated by an elevator both on the way up and down. The steps to create zero gravity inside the elevator were described in the previous chapter. Some human engineering design considerations must be taken into account in order for the elevator to be an operable zero G system.

High G is defined as greater than our weight we experience everyday. A trained fighter pilot can endure up to 9G's. That is to say he/she can suffer 9 times of his/her weight on him/her. An untrained person can reasonably take 4G's with a right posture and preparation. It is an exciting sensation for passengers to experience suddenly shifting from a high G state to weightlessness. But the transition from zero G to a high G state must be gradual for safety reasons. It is fine and fun to let the elevator go free climbing at the injection point. However, coming back from 0 G to high G must be gradual at the recovery point. This can be achieved by controlling the elevator with a constant jerk as defined above. The constant jerk means the acceleration or G force will be increased gradually from recovery point to the bottom and then up to the next injection point in the next cycle of weightlessness ride.

In simulating of reduced gravities, some questions are: How should the elevator move down so that the inside passenger will experience 1/6 of G as on the surface of the Moon?

How should the elevator move up so that the inside passenger will experience 1/6 of G as on the surface of the Moon?

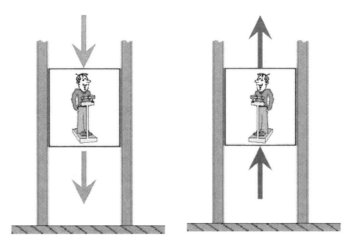

Downward acceleration   Upward "Deceleration"

If the reader is not careful, he/she might think the first question is straightforward and assume the elevator needs to be moving down with 1/6 of g to achieve the desired effect. Then think again.

To serve the purpose of simulating a reduced gravity, the elevator must "absorb" part of the gravitational acceleration g on Earth's surface so that the "residual" acceleration can be used to simulate desired reduced gravities inside the elevator. The method of simulating the gravity on the Moon should be as following:
Since the gravitational acceleration ratio P on the surface of the Moon is 1/6 or 16% of the Earth, the elevator system must be falling down at an acceleration 1-P = 5/6 of the Earth's gravitational acceleration g in order to simulate 1/6 g inside the elevator.

The second question is a bit trickier because an initial condition is presumed. A sufficient initial upward speed is needed for the elevator to slow down after reaching the injection point on the way up. The principle is similar to the first question. Now we want to find the deceleration. A deceleration of 1-P = 5/6 of g is needed for the elevator to have a 1/6 of residual g inside.

How to simulate variable gravities on a parabolic track

This is a somewhat more complicated problem because it involves horizontal velocity. The motion is not straight up and down as the elevator questions. The following description also introduces sets of seemingly complicated equations. Readers should not be turned off by the math and concepts because all the physics and math manipulation used in the following discussions are covered in high school curriculum. Without the misconceptions of zero gravity, the solution is just obtained by combining different applications of few simple principles in Newtonian mechanics.

For simplicity, we use scientific or metric measurements and the approximation of $g = 10$ m/sec^2, the gravitational acceleration on the surface of the Earth.

Parabolic equation for the rail routes can be derived as following:
Given a constant horizontal velocity v_0, the horizontal distance x traveled in time t will be

$x = v_0 t$

Vertical distance traveled in time t will be

$y = -1/2\ g\ t^2$, where minus sign is to indicate the downward direction.

By eliminating t, we get

$$y = -g/(2v_0^2) \cdot x^2 \quad \textbf{equation (1)}$$

The parabolic route is governed by this equation,
where
x is the horizontal coordinate, y is the vertical coordinate in meters.
g is the gravitational acceleration near the surface of the Earth, the value is
$g = -9.8 \text{ m/sec}^2$
v_0 is the horizontal component of the velocity, v_0 remains constant for a particular parabola

The derivative of equation (1) with respect to x is
$y' = dy/dx = -(g/v_0^2) \cdot x \quad \textbf{equation (2)}$
y' is the slope of the parabola at any point x.

Since $v_y = dy/dt$, $v_x = dx/dt$, we have
$$v_y/v_x = dy/dx = -(g/v_0^2) \cdot x \quad \textbf{equation (3)}$$

The vehicle velocity at each point on the path to achieve 0 G and reduced G can be derived as follows:

For zero G:

$y' = v_y/v_x = -(g/v_0^2) \cdot x$ where horizontal speed remains constant $v_x = v_0$
thus $v_y = -(g/v_0) \cdot x$
Each parabola has a unique v_0 horizontal velocity. Thus by vector adding x- and y- components of the velocity, we get the vehicle velocity on rail at any point x for producing weightlessness.
$v = v_x + v_x$
Magnitude of $v = |v| = (v_0^2 + (g^2/v_0^2) \cdot x^2)^{1/2}$

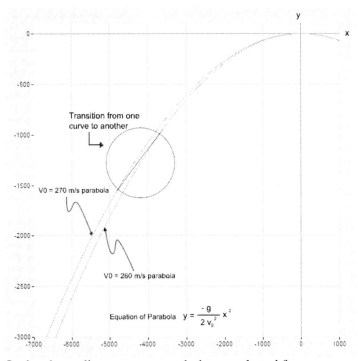

In the above diagram, two parabolas are plotted for different horizontal speeds. The total time period for 0 G in this example can be 49 seconds for the height is 3,000 meters. Note that the horizontal and vertical scales are different for the resolution of the graph. A scaled version of a similar parabola is in the next diagram.

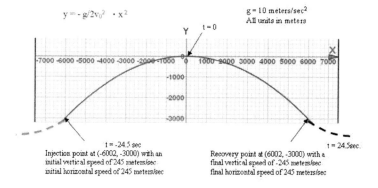

A parabola with horizontal speed v_0 = 245 meters/sec

For reduced Gs:

Commercial zero G flights can simulate the reduced gravities of walking on the Moon or Mars by taking different parabolic flight paths as we mentioned before. The proposed zero G high speed train will be on a fixed parabolic track. It is impossible for the land vehicle to do the same? We are now ready to attack this problem.

The system has to "absorb" part of the gravitational acceleration in order to leave the so-called "residual" gravitational acceleration inside the cabin as the reduced gravity. For example, for the lunar to earth gravity ratio $p = 1/6 \sim 16\%$, the vehicle has to be "falling downwards" with $(1-p)g$ meters/sec^2, where g is approximately 9.8 meters/sec^2.

Now let $v_h = v_0 (1-p)^{1/2}$, **equation (4)**

where v_0 is the horizontal velocity for 0G

and $v_y = -g((1-p)/v_h)x$, **equation (5)**; Please note that v_h is a constant and v_y is not. v_y is a variable with x, the horizontal distance.

We will prove that $y = -\frac{1}{2}(1-p)gt^2$.
That is to say that we achieve the reduced gravity $(1-p)g$

$$\begin{aligned}
y &= \int v_y dt \\
&= \int -g((1-p)/v_h)x\, dt \quad \text{from equation (4)} \\
&= \int -g((1-p)/v_h)v_h t\, dt \quad x = v_h t \\
&= \int -g(1-p)t\, dt \\
&= -\tfrac{1}{2}g(1-p)t^2
\end{aligned}$$

We can look it at another way. Let v_h represent the required horizontal velocity for achieving a reduced gravity $(1-p)g$. Equation (1) can not be changed because the parabola track path is fixed. That is to say the coefficient $-g/(2v_0^2)$ is also fixed. Multiplying both the denominator and numerator of the coefficient by the factor of $(1-p)$, we get
$-g(1-p)/(2v_0^2(1-p))$

Comparing this with the coefficient of Equation (1), $v_0^2(1-p)$ is now part of the coefficient that needs to be the squareroot of the new horizontal velocity.
Thus, we get
$v_h^2 = v_0^2(1-p)$ **equation (6)**
and
$v_h = v_0(1-p)^{1/2}$, where v_0 is the horizontal velocity for 0G

The vertical component of the velocity v_y can be derived from the derivative equation (2),
$$y' = v_y / v_h = - g/ v_0^2 \cdot x \quad \text{equation (2)}$$

We know from equation 6 above $v_h^2 = v_0^2(1-p)$, therefore $v_0^2 = v_h^2/(1-p)$

By replacing v_0^2 in equation 2, we get
$$v_y = (- g/ v_0^2 \cdot x) v_h$$
$$= - (g(1-p) / v_h) \cdot x$$

Or $v_y = \dfrac{-g(1-p)}{v_h} \cdot x$

v_y is the vertical component of the vehicle velocity at each point x along the x-axis.

Again by the sum of x- and y- components of the velocity, we get the vehicle velocity on the rail at any point x for simulating a reduced gravity.

$$v = (v_h^2 + v_y^2)^{1/2} = (v_0^2(1-p) + (g^2(1-p)^2 / v_h^2) \cdot x^2)^{1/2}$$
since $v_h^2 = v_0^2(1-p)$
$$v = (v_0^2(1-p) + (g^2(1-p) / v_0^2) \cdot x^2)^{1/2}$$

Thus any reduced gravity condition can be simulated on the fixed track path by adjusting v accordingly.

For lunar gravity, 16% or ~1/6 of Earth's gravity is created in the cabin.
The system must use up or "absorb" 1- p = 5/6 of the Earth's gravitational acceleration so that the passenger can experience the 1/6 of Earth's gravity (gravity on Moon's surface). Suppose we have a parabola track with $v_0 = 270$ m/s.

$v_h = v_0 (1-p)^{1/2} = 246$ m/sec

$v_y = -(g(1-p)/v_h) \cdot x = -0.0339 x$ m/sec

For Martian gravity, 38% or 3/8 of Earth's gravity is to be made in the cabin.
The system must use up or "absorb" $1-p = 5/8$ of the Earth's gravitational acceleration so that the passenger can experience the 38% of Earth's gravity (gravity on Mars). Suppose the vehicle is on a $v_0 = 245$m/s parabola route.

$v_h = v_0 (1-p)^{1/2} = 193.7$ m/sec
$v_y = -g(1-p)/v_h \cdot x = -0.03226 x$ m/sec

Again, the minus sign here is to indicate that the direction of vertical acceleration is always pointing downward. The system can be operated as an intercity transit. A half G vehicle with a horizontal velocity v_h of 191meters/sec on a $v_0 = 270$ parabola track path can be reasonably used as a passenger transit.

Interesting facts used by zero gravity

$g = 9.81$ m/s², the gravitational acceleration near the surface of the Earth, is a measurable quantity. We had been told that we weigh only 1/6 of our weight on Moon, and 3/8 of our weight on Mars. This is to say the gravitational acceleration on the Moon is 1.63 m/s^2 or on the Mars 3.67 m/s^2. But how can we tell the gravitational accelerations of Moon and Mars way before anyone set foot on other heavenly bodies? The readers may also wonder how we can find out the mass of the Earth and other heavenly bodies? These questions can be solved with Newton's law of universal gravitation and three laws of motion. Newton's equation of universal gravitation had a mystery G, the universal gravitation constant. The value of G could not be determined during Newton's time. Not until nearly a century later (1798) did Lord Henry Cavendish determine its value with his brilliant experiment. Once the value of G was known, all the above questions could be solved. The value of G is an extremely small numerical value. It accounts for the fact that the force of gravitational attraction is only appreciable for an object with large mass. While two people will indeed exert gravitational pulls upon each other, the forces are too small to be noticeable. Only the gravitational force between a person and the planet becomes noticeable. That is the person's weight. The readers can easily imagine that the numerical values of the constants for magnetic and electric attraction forces will be much larger. With the known value of the upper case G, the lower case g on the Moon or Mars can be determined easily. It is not in the scope of this book to go through the details of Newton's laws. Readers can find the information from various sources if interested.

Center of mass

The Earth is a huge sphere with its mass unevenly distributed inside its volume. If we want to calculate the gravitational attraction force between the Earth and an object on Earth, the calculation would be daunting if not impossible without the concept of center of mass. Center of mass is the point in a system of bodies at which the mass of the system may be considered to be concentrated and at which external forces may be considered to be applied. This concept was developed by Newton and proven by yet another one of his invention: calculus.

Conic Sections

Conic sections have been studied for over 2000 years. They are a set of curves can be derived by cutting a cone through different angles. Circles, ellipses, parabolas and hyperbolas are very useful in astronomy to describe the motions of heavenly bodies. Parabolas are the foundation for artillery and missile trajectory analysis in the past. The latest smart weapons use Laser or Global Positioning System GPS to guide the projectiles.

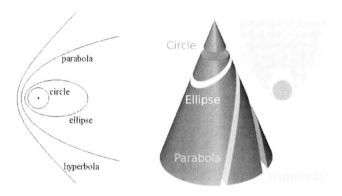

Are They Really Parabolic Paths?

The aerospace industry and business have being using the term parabolic flight since the invention of the zero G weightlessness flight. For all practical purposes, the path is a parabola. Mathematically, it is actually a small upper portion of an extremely flattened ellipse similar in shape as the orbit of a comet around the sun. The ellipse has one of its foci at center of mass of the Earth. The ellipse passes through the inner solid body of the Earth and forms a perfect mathematical curve of a conic section. It is due to the fact that we can imagine that all of the Earth's mass is concentrated at one point at center of mass deep inside the core.

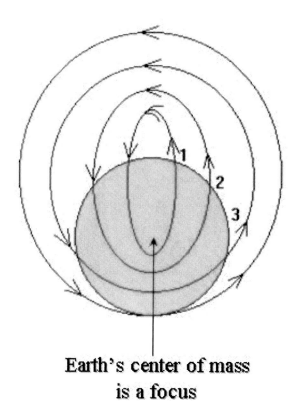

Earth's center of mass is a focus

Appendix 1   Glossary

Acceleration   is the rate of change of velocity. It is caused by applying a force on an object.

Acting-at-a-Distance or **Non-Contact Forces** Gravitational, Magnetic, Electricity attractions are forces acting on objects without being in direct contact with them. These forces obey the inverse square rule. An example is Newton's law of universal gravitation.

Artillery can project ammunitios far beyond the range of the personal weapons or small firearms.

Burj Khalifa Building is a skyscraper in Dubai, United Arab Emirates, and the tallest man-made structure ever built, at 828 m (2,717 ft).

Catapult is a device used to throw or hurl a projectile a great distance at high speeds without the aid of explosive devices.

Contact Force In contrast to acting-at-a-distance forces, contact forces act on objects as a result of direct contact.

Force   a **force** is any influence that causes a free body to undergo an acceleration in physics

Free Climb or Ascension   Not to be mixed with the sport of free climbing. It means an object moving up with a given initial speed and slowing down by the gravity without the influence of any contact force like wind effects.

Free Fall   A free falling object is an object which is falling under the sole influence of gravity by ignoring the effects of air. A free falling object on Earth accelerates downwards at a rate of 9.8 m/s/s.

High G or Hyper Gravity is caused by acceleration greater than the gravitational acceleration g.

High Speed Rail HSR   There are two types of High Speed Rail technologies, conventional wheeled HSR and magnetic levitation HSR.

Injection Point is the transition point from a high G to weightlessness state or from an accelerated projection to the free flying state. The injection point does not need to be coincided with the recovery point for the 0 G elevator. The injection point does not need to be symmetric to the recovery point for the 0 G parabolic path either.

Jerk, jolt or surge is the rate of change of acceleration.

Kamikaze   神風, Japanese: "divine wind" is the name of suicide aircraft attacks by military pilots of the Empire of Japan against American fleets or B-29 formations during the final stages of the Pacific campaign of WW II.

Linear Electromagnetic Motor is an induction motor producing a drive along a straight line rather than a torque of rotation.

Maglev or Magnetic Levitation is a high speed rail technology to eliminate friction by floating on the path track.

Microgravity is zero gravity or weightlessness because it is hard to obtain a perfect zero G in many applications.

Ordnance is the collection of weapons and ammunitions beyond the small personal firearms. Not to be mixed with ordinance

Parabola is the curve for a projectile in an ideal condition, a curve from the family of conic sections.

Pitch is the nose up or down of an aircraft, rotation around Y-axis in the diagram.

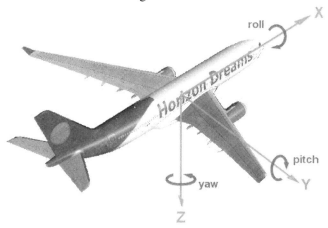

Pressure is the force applied to a surface area.

Principle of Equivalence   an observer or instrument like accelerometer can not tell the difference between gravitational acceleration and acceleration caused by an applied force. Einstein's theory asserts that the gravitational "force" as experienced by us is the same as the force experienced by an observer in an accelerated vehicle.

Recovery Point is the transition point from weightlessness to a high G. For the zero G flight is the point where the aircraft is pulling up from a zero G dive. For the zero G elevator is the point where the elevator starts to brake or slow down. The recovery point does not need to be coincided with nor symmetric to the injection point.

Reduced Gravity is the gravitational acceleration less than 1 G, larger than 0 G.

Speed is the quantity of velocity without considering the direction.

Surge   See Jerk, Jolt

Vacuum is void of air

Variable is a letter that representing a changeable value in a mathematical expression.

Vector is a physical quantity having both magnitude and direction like velocity and acceleration

Weight   the weight of an object is the force exerted on the object by gravity.

Weightlessness is used interchangeably with Zero Gravity or zero G.

Appendix 2   Circular Motions and Artificial Gravities

The centripetal acceleration of a uniform circular motion is

$$a = \frac{v^2}{r}.$$

Where v is the constant tangential speed of the circular motion and r is the radius of the circular motion.

The acceleration is due to an inward-acting force, which is known as the centripetal force (meaning "center-seeking force"). It is the force that keeps an object in uniform circular motion. From Newton's Second Law of Motions, the centripetal force $\mathbf{F}_c$ for an object in uniform circular motion is related to the object's acceleration by

$$\mathbf{F}_c = m\mathbf{a},$$

where *m* is the mass of the object.

Since the magnitude of the acceleration is given by $a = v^2/r$, the magnitude of the centripetal force is given by

$$F_c = m\frac{v^2}{r}.$$

The centripetal force can be provided by many different things, such as tension (as in a rope or sling), friction (as between tires and road for a turning car), or gravity (as between the Sun and the Earth). In our case of the artificial gravities, the centripetal force is provided by the spacecraft. The variable artificial gravities are adjustable with the rotational speed.

The centrifugal force (meaning "center-fleeing force") is sometimes called a fictitious force because an outside observer in an inertial frame will find that the person inside the spinning donut-shaped spacecraft is only obeying Newton's First Law of Motion, but pushed by the outer edge of the spacecraft toward the center of the circular motion. See the following diagram.

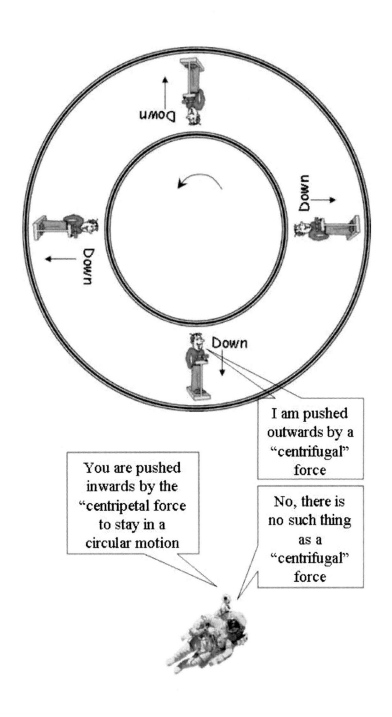

Appendix 3   References:

Advanced Airway Management: International Space University

German Aerospace Center:
http://www.dlr.de/en/desktopdefault.aspx/tabid-734/1210_read-15591

Encyclopædia Britannica

Natural and Applied Sciences,
University of Wisconsin - Green Bay

Physics Classroom
http://www.physicsclassroom.com/Class/circles/

USA today
http://www.usatoday.com/news/graphics/zerog/flash.htm

Virgin Galactic

Wikipedia

CPSIA information can be obtained at www.ICGtesting.com
Printed in the USA
BVOW02s2204030416

442827BV00019B/104/P